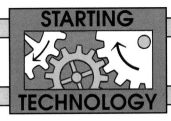

ELECTRICITY

John Williams

**Illustrated by
Malcolm S. Walker**

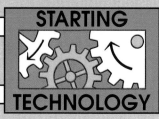

Titles in this series

AIR
COLOUR AND LIGHT
ELECTRICITY
FLIGHT
MACHINES
TIME
WATER
WHEELS

Words printed in **bold** appear in the glossary on page 30

© Copyright 1991 Wayland (Publishers) Ltd

First published in 1991 by
Wayland (Publishers) Ltd
61 Western Road, Hove
East Sussex BN3 1JD, England

Editor: Anna Girling
Designer: Kudos Design Services

British Library Cataloguing in Publication Data
Williams, John
  Electricity.
  1. Electricity
  I. Title II. Walker, Malcolm 1947- III. Series
  537
  ISBN 0 7502 0169 X

Typeset by Kudos Editorial and Design Services, Sussex, England
Printed in Italy by Rotolito Lombarda S.p.A.
Bound in Belgium by Casterman S.A.

CONTENTS

It's Electric! 4
Lighting Up 6
Switch On! 8
Natural Electrics 10
Conductors 12
Electrical Toys 14
Playing a Game 16
Messages 18
Magnets 20
Printed Circuits 22
Choice Cards 24
Electric Buggies 26
Notes for Parents and Teachers 28

Glossary 30
Books to Read 31
Index 32

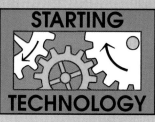

IT'S ELECTRIC!

Nowadays we cannot do without electricity. Look around you. Many everyday things work by electricity. Lights, heating, trains, telephones, computers, even parts of cars, all need electricity to work. Think what we would do without it!

These are electricity pylons. The thick metal cables carry electricity away from the power stations where it is made.

Looking at batteries and bulbs

WARNING: All the models in this book use small batteries and bulbs. Never play with the mains electricity in your home or at school. It is very dangerous.

1. Make a collection of as many different **batteries** as you can find. Are they all the same shape? Are some heavier than others? Are they the same at each end?

2. Look at the marks on the batteries. Look for a plus sign (+) and a minus sign (–), and any numbers before a letter 'V'. Do all the batteries have these marks? How many different numbers can you see?

3. Look closely at some small **bulbs** — the kind used in torches. Use a **magnifying glass** if you have one. What can you see inside the bulb? Can you see any numbers before a letter 'V' on the bulbs? Are the numbers always the same?

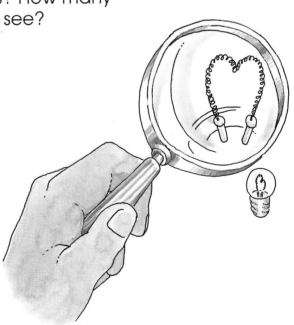

LIGHTING UP

Making circuits

You will need:

Two 3.5V bulbs
Two bulb holders
Several pieces of plastic-
 covered wire
A 4.5V battery

1. The wires should have the inner strands of metal showing at both ends. Fix wires to the two screws on a bulb holder.

2. Fix the other ends of the wires to the two metal flaps, or **terminals**, on the battery. It helps to use paperclips or crocodile clips. When the bulb lights up you have made an **electric circuit**.

3. Now, using another piece of wire, make a circuit which has two bulbs in it. This is called a **series circuit**. Do both bulbs shine with the same brightness?

4. Unscrew one bulb. What happens to the other bulb?

More circuits

Make up more circuits with two bulbs. Try to make a circuit where you can unscrew one bulb, but the other stays lit up. This is called a **parallel circuit**.

Look at the beautiful lights on this Christmas tree. Do you think it is better to use series circuits or parallel circuits to make sure the bulbs stay alight?

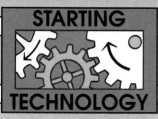

SWITCH ON!

Making switches

You will need:

Thin card
Kitchen foil
Paper fasteners
Glue
A matchbox
Paperclips
Wire, 4.5V battery, bulb and holder

Type one

1. Cut out a piece of card about 10 cm square. Fold it in half.

2. In the middle of each half, glue a small piece of kitchen foil, about 2 cm square.

3. Push a paper fastener through each piece of foil. The tops of the two fasteners should touch each other when the card is closed.

4. Attach wires to the wings of the fasteners and connect them to the bulb and battery to make a circuit. When you close the card the bulb will light up.

Type two

1. Push two paper fasteners through the top of the matchbox. Make sure they do not touch.

2. Attach a paperclip to one of them. Make sure it is big enough to reach the other fastener.

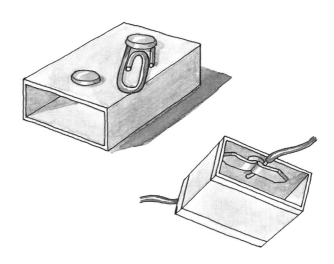

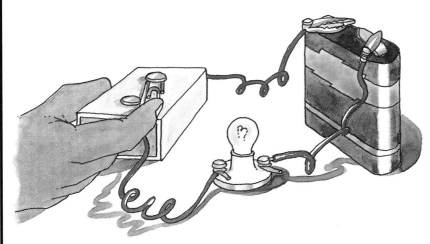

3. Fix wires to the wings of each fastener and connect them to your circuit.

4. Swivel the paperclip to make the bulb light up or go out.

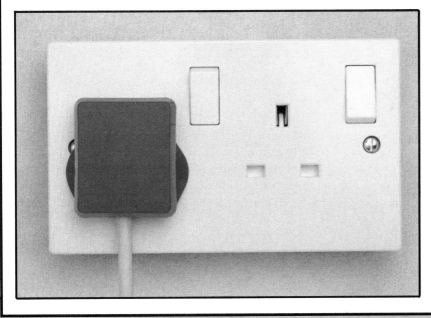

A plug point like this in your home or at school has switches on it. They work in a similar way to the switches you can make.

NATURAL ELECTRICS

People have known about electricity for thousands of years. They have always been fascinated by natural forms of electricity, like **lightning**. Electricity is all around us. Here are some things you can do to find it!

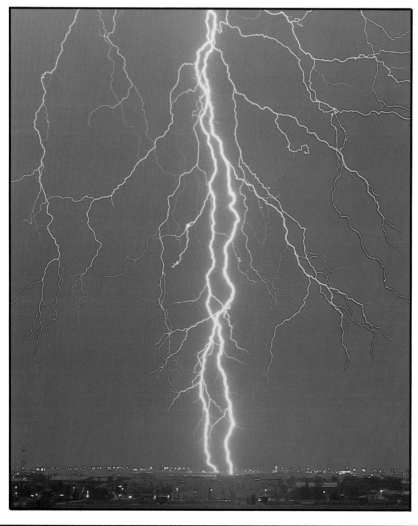

This flash of lightning is like a giant spark of electricity stretching from the clouds in the sky to the ground.

Finding natural electricity

You will need:

A plastic ruler
Cotton cloth
Tissue paper
Cotton thread
A pencil
A balloon

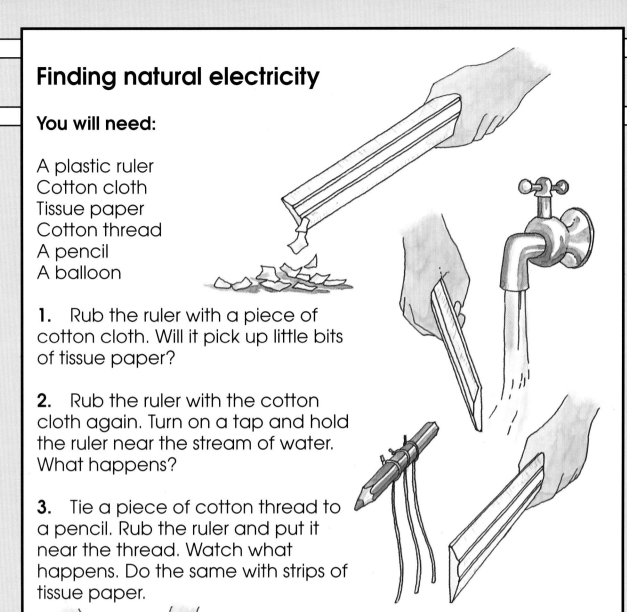

1. Rub the ruler with a piece of cotton cloth. Will it pick up little bits of tissue paper?

2. Rub the ruler with the cotton cloth again. Turn on a tap and hold the ruler near the stream of water. What happens?

3. Tie a piece of cotton thread to a pencil. Rub the ruler and put it near the thread. Watch what happens. Do the same with strips of tissue paper.

4. Blow up the balloon. Rub it with cotton or wool cloth. Can you make it 'stick' to a wall?

This kind of electricity is called static electricity. Static means to stay still.

Now think again about the electricity you used to make a bulb light up. This electricity did not stay still. It moved through the wires and bulbs around a circuit. This is called current electricity.

11

CONDUCTORS

Electricity moves more easily through some **materials** than others. If it can move easily, the material is called a good **conductor**. Metals are good conductors. Rubber and plastic are bad conductors.

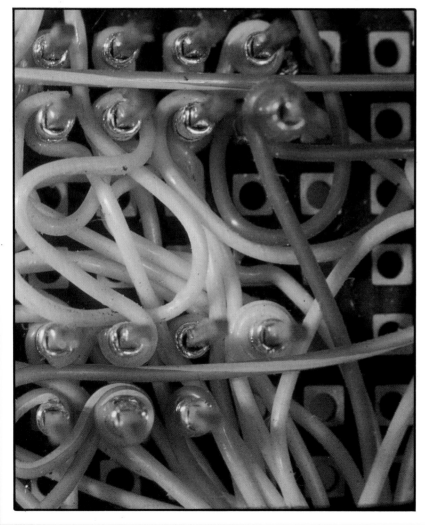

These wires are covered in plastic to stop the electricity moving from one wire to another. Plastic is a bad conductor.

Testing for conductors

You will need:

Card
Scissors
Paperclips
Wire, 4.5V battery,
  bulb and holder

Objects to be tested,
  such as wood, paper,
  string, metal scissors,
  plastic, pencil leads

1. Cut a piece of card about 15 cm long and 5 cm wide. Make two slits in it, about 3 cm from each end.

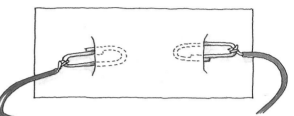

2. Attach wires to two paperclips. Slide the paperclips into the slits. Connect the wires to the bulb and battery to make the rest of the circuit.

3. Place each object to be tested across the space between the two paperclips. If the light shines, then electricity is going through the material.

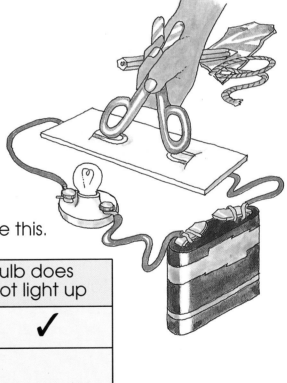

4. Make a list of the materials like this.

MATERIAL	Bulb lights up	Bulb does not light up
Wood		✓
Metal scissors	✓	
String		✓

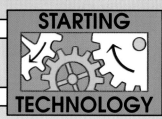

ELECTRICAL TOYS

Making robot eyes

You will need:

Cardboard boxes
Card
Kitchen foil
Wire, 4.5V battery, two bulbs and holders
Paper fasteners
Glue

1. Make a robot out of boxes, card and kitchen foil.

2. Wire up your bulbs in a parallel circuit. Make a switch from card and paper fasteners and include it at one end of your circuit.

3. Fix the light bulb eyes in place on your robot. Glue the switch to one side of the head to look like an ear. Make a matching ear for the other side. When you close the switch the eyes will light up.

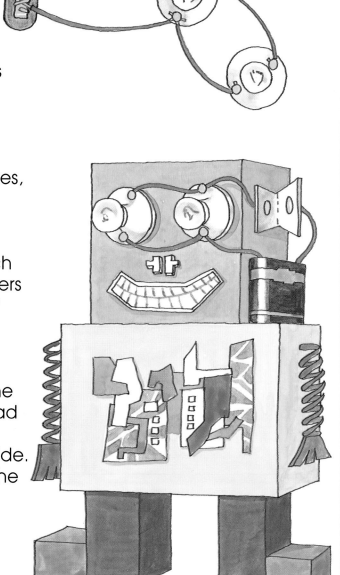

Making choice cards

You will need:

Card
Paper fasteners
Dowel rods
Sticky tape
Wire, 4.5V battery,
 bulb and holder

1. Push two rows of paper fasteners into a large piece of card.

2. Make up some questions and answers. You could ask the names of shapes or animals. Write the questions by the left-hand row of fasteners and the answers by the right-hand ones. Make sure they are muddled up.

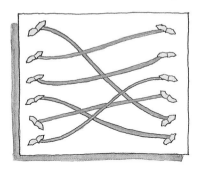

3. Turn the card over and use wires to connect up the fasteners, so that each question is linked to its correct answer.

4. Wire the bulb and battery together. Connect a wire to the other terminal on the battery and attach the other end to a piece of dowel with sticky tape. Do the same with a wire from the bulb.

5. Answer the questions on the card by touching two paper fasteners with the wires from the bulb and battery. If you answer the question correctly the bulb will light up.

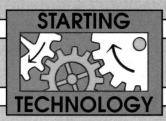

PLAYING A GAME

A steady hand

You will need:

A plank of wood about 60 cm long and 20 cm wide
Two cotton reels
A short piece of dowel
Plasticine
Glue

Thick uncovered wire, such as a metal coat hanger
Wire, 4.5V battery, bulb and holder
Sticky tape

1. Fix the cotton reels to the plank with glue. Fill the centres of the reels with plasticine.

2. Bend the thick wire into a zigzag shape and stick the ends into the plasticine in the reels.

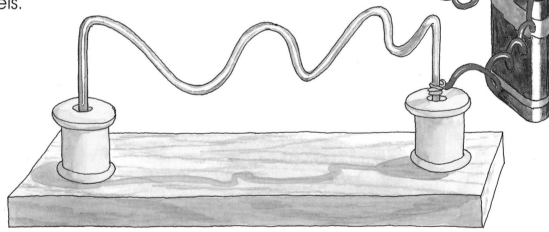

3. Connect the bulb and battery together and then wire one end of the circuit to the zigzag wire, near one of the reels.

4. Take a wire from the other end of the circuit and bend it into a loop around the zigzag wire. Tape it to a piece of dowel to keep it straight.

5. When the loop touches the zigzag wire the bulb will light up. Try to move the loop all the way along without touching the zigzag.

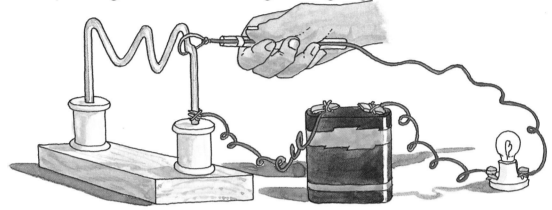

Some of these bright, colourful lights flash on and off.

17

STARTING TECHNOLOGY

MESSAGES

Morse code is a special way of sending messages using long and short flashes of light or buzzes of sound. You can make up your own code and make an electrical 'tapper' to send messages to your friends. Then look up Morse code in a book. Is it like the code you made up?

This man is sending out a message in Morse code. He is using the tapper with his right hand.

Making a code tapper

You will need:

A block of wood, 12 cm long, 8 cm wide and 2 cm deep
A stick of wood, 8 cm long and 1 cm square
A short piece of dowel
Wire, 4.5V battery, bulb and holder
Small nails
Rubber bands
Drawing pins
Glue

1. Glue the dowel across the middle of the block of wood.

2. Push a drawing pin into the block, about 1 cm from the edge. Join a wire from the drawing pin to the bulb.

3. Place the wooden stick across the dowel so that one end is over the drawing pin. Put a rubber band round the other end and round the block of wood. Push two nails into the block, one on each side of the stick, to keep it straight.

4. Push a drawing pin into the stick of wood, making sure that the two drawing pins touch when the stick is pushed down.

5. Fix a wire to the second drawing pin and connect up the bulb and battery to make a circuit. When the stick is pushed down, and the pins touch, the bulb will light up and you can tap out a message.

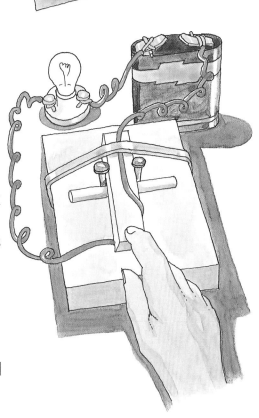

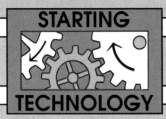

MAGNETS

Less than 200 years ago, a scientist called Michael Faraday discovered how to use **magnets** and copper wire to make electricity. He found that electricity and magnetism are very closely linked. Magnets can be used to make electricity and electricity can be used to make magnets.

Magnets come in different shapes. Many are shaped as bars or horseshoes.

Making magnets with electricity

You will need:

Iron nails
Wire, 4.5V battery,
 bulb and holder
Paperclips

1. Very carefully, wind a length of wire around an iron nail. Make sure the wire is covered with plastic, except at the ends. These ends must not touch the nail.

2. Connect up your coiled wire into a circuit with the bulb and battery. The bulb will tell you if the circuit is working properly.

3. See how many paperclips the nail will pick up.

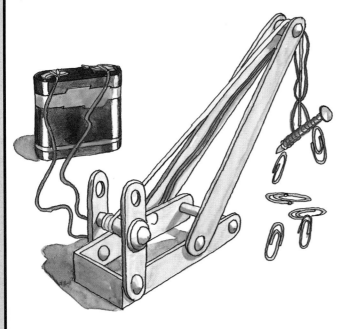

4. Make other **electromagnets**. Wind two layers of wire around a nail, or use two nails together. Do these make better magnets?

5. Make a large crane out of a construction kit. Use one of your electromagnets instead of a hook to pick things up.

STARTING TECHNOLOGY

PRINTED CIRCUITS

The circuits you have made so far have used wires to connect them up. Many electrical items, such as computers or washing machines, would need masses of wires to make them work. The circuits would be very messy. Instead, they use **printed circuits**.

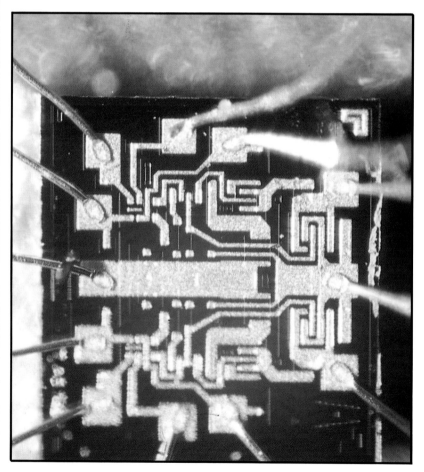

This picture shows a tiny printed circuit enlarged many times. Some circuits, called micro-circuits, are not much bigger than the full stop at the end of this sentence.

Making printed circuits

You will need:

Card
Kitchen foil
Scissors
Glue
Wire, 4.5V battery,
  bulbs and holders

1. Cut two strips of kitchen foil and stick them to a piece of card, about 30 cm by 20 cm. The strips do not have to be straight.

2. Connect the strips at one end by wiring them to a bulb.

3. Attach wires to the other ends of the strips and connect them to the battery.

4. You can make a switch by cutting a flap in one of the foil strips.

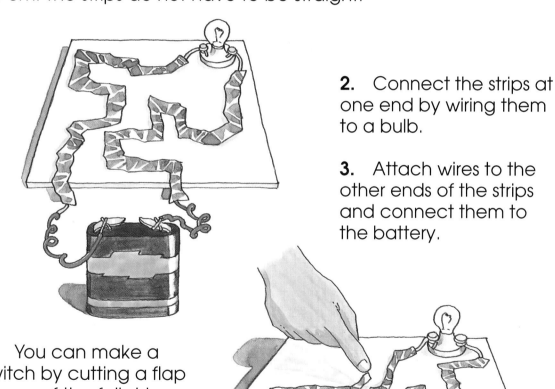

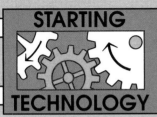

CHOICE CARDS

Making a choice card with a printed circuit

You will need:

Card
A pencil and ruler
Kitchen foil
Paper fasteners
Scissors
A stapler
Wire, 4.5V battery,
  three bulbs and holders

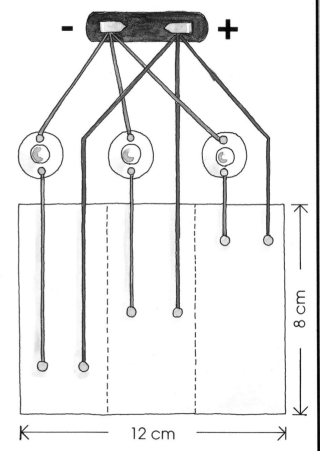

1. Cut a piece of card 12 cm long and 8 cm wide. Draw lines to divide it into three parts.

2. Push six paper fasteners through the card, in the places shown in this drawing.

3. Attach wires to the fasteners and connect each pair of fasteners to a bulb and the battery.

4. Staple a second piece of card over the first, stapling along the sides and along your dividing lines. Leave the top and bottom open.

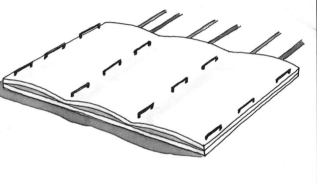

5. Cut pieces of card 8 cm long and almost 4 cm wide. On one side stick short strips of kitchen foil. They must be in exactly the right places to match the fasteners on the big piece of card.

6. Turn the strips of card over and draw pictures on them. You could draw different animals or flowers.

7. Write the name of each picture on the matching slot on the large card. When the correct card is pushed into the slot, the foil will touch the fasteners and the bulb will light up.

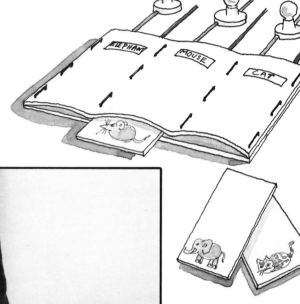

Your computer at school probably has a printed circuit board very similar to this one.

ELECTRIC BUGGIES

Making an electric-powered buggy

You will need:

Two sticks of wood, 1 cm square and 30 cm long
A piece of balsawood, 30 cm long and 6 cm wide
Two pieces of dowel, about 35 cm long
Two plastic bottles
Wood glue
Rubber bands
A small electric motor (from a toy or model shop)
Wire, 4.5V battery
Sticky tape

1. Make a hole in the bottom of each plastic bottle. Push the pieces of dowel through the hole and out through the neck of each bottle. The bottles should spin round easily. Put two rubber bands round each bottle to act as grippers.

2. Fix the dowels to the square sticks of wood with rubber bands. Glue the balsawood to the sticks, between the bottles.

3. Fix the motor to the balsawood with sticky tape. Stretch one of the rubber bands you put round the bottles to the **spindle** on the motor.

4. Wire the motor to the battery and fix the battery to your buggy with sticky tape.

5. Test your buggy on different surfaces, rough and smooth. Will it go up slopes?

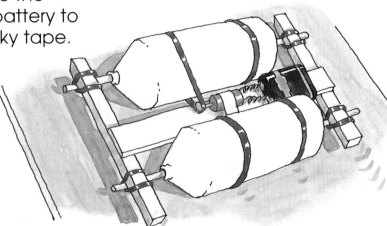

This boy's lorry has a remote control box. You could make a switch for your buggy, attached to long wires, to use as a remote control.

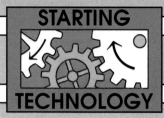

Notes for Parents and Teachers

SCIENCE

Children should be allowed to play with bulbs and batteries before they are given any structured guidance. They can then be introduced to more formal circuits. After this, it is important that they should use their knowledge to make working models.

Children will often ask what the various marks and signs on the batteries and bulbs mean. The volt sign can be explained as the force, or push, of an electric current. It should be noticed that this has nothing to do with the size of the battery. The + and – signs are more difficult. However, if the wires on an electric motor are reversed then the motor will rotate the opposite way. Through this, children will obtain an insight into why there are two terminals on the battery, and how they are different.

TECHNOLOGY

Children should be encouraged to develop, test and improve the models that they make to gain an understanding of the technological design process. They can make simple designs as their models progress.

Control technology can easily be incorporated into work on electricity. Children can wire up various models to a computer, through a suitable control box. Using a simple program, children can order the computer to carry out all kinds of tasks – for example, making a light flash in sequence.

All the work in this book has been carried out by young children. Only simple tools are needed, such as screwdrivers, scissors, junior hacksaws and wire strippers.

LANGUAGE

Any science or technology topic will provide opportunities for children to extend their language skills. The topic of electricity should give rise to much discussion which can lead to interesting creative and factual writing.

MATHEMATICS

Where necessary, children should be encouraged to make careful measurements, both of the materials used and in the testing of their finished models.

National Curriculum Attainment Targets

This book is relevant to the following Attainment Targets in the National Curriculum for **science**:

Attainment Target 1 (Exploration of science)   The designing, testing and making of the various models answer many of the requirements in this Attainment Target.

Attainment Target 11 (Electricity and magnetism)   All the work in this book is relevant.

The following Attainment Targets are included to a lesser extent:

Attainment Target 6 (Types and uses of materials)   The testing of materials as conductors is relevant.

Attainment Target 10 (Forces)   Work on magnets involves the forces of pushing and pulling.

Attainment Target 13 (Energy)   Electricity is a form of energy.

This book is relevant to the following Attainment Targets in the National Curriculum for **technology**:

Attainment Target 1 (Identifying needs and opportunities)   Children will observe the uses of electricity – for example for light, movement and magnetism. By working with circuits and switches, they will observe how these devices may be employed and should be encouraged to suggest further uses.

Attainment Target 2 (Generating a design)   Children should look at the various designs for switches, circuits and printed circuits and then decide how best to proceed. They should propose their own designs through drawings and later review them to see how they could be improved.

Attainment Target 3 (Planning and making)   All the projects in this book require children to choose and use suitable materials and tools.

Attainment Target 4 (Evaluating)   Children should judge how well their circuits, switches, toys, magnets and buggies have worked.

Teachers should also be aware of the Attainment Targets covered in other National Curriculum documents – that is, those for language, mathematics and history.

GLOSSARY

Battery A container with special chemicals in it that produce electricity.

Bulb A glass object, shaped like a pear, which has a thin metal wire in it. When electricity passes through the wire it gets hot and gives out light.

Conductor Something that electricity will move through.

Dowel A wooden rod.

Electric circuit A loop of wires and objects connected up so that electricity will flow round it.

Electromagnet A magnet made by coiling a wire round a piece of iron or steel and passing an electric current through the wire.

Lightning An electrical flash of light in the sky during a thunderstorm.

Magnet An object that can pull things made of iron towards it.

Magnifying glass A curved piece of clear plastic or glass that makes things look bigger.

Materials Substances from which other things can be made.

Parallel circuit An electric circuit where several objects, such as light bulbs, are connected up side by side so that the same amount of electricity goes through all of them.

Printed circuit An electric circuit which is printed in metal on a board.

Series circuit An electric circuit where several objects, such as light bulbs, are connected up one after another so that electricity goes through each one in turn.

Spindle A thin rod or pin.

Terminals The metal flaps or buttons on a battery which you attach wires to when you are connecting up an electric circuit.

BOOKS TO READ

Bulbs and Batteries by Ed Catherall (Wayland, 1986)
Electricity by John and Janet Clemence (Macdonald, 1987)
Electricity and Magnetism by Kay Davies and Wendy Oldfield (Wayland, 1991)
Electricity and Magnetism by Kathryn Whyman (Franklin Watts, 1986)
Fun with Magnets by Ed Catherall (Wayland, 1985)
Switch on a Light by Joy Richardson (Hamish Hamilton, 1988)

Picture acknowledgements
The publishers would like to thank the following for allowing their photographs to be reproduced in this book: Eye Ubiquitous 9 (Paul Seheult); Science Photo Library 4 (David Parker), 12 (Ray Simon), 20 (Richard Megna); Telefocus 18; Tony Stone Worldwide 10, 17 (Tim Brown); Tim Woodcock 27; Zefa 7, 22 (G. Mabbs), 25. Cover photography by Zul Mukhida.

INDEX

bar magnets   20
batteries   5
buggies, electric   26-7
bulbs   5

choice cards   15, 24-5
circuits   6-7, 11, 22-3
code tappers   18-19
computers   4, 22, 25
conductors   12-13
current electricity   11

electromagnets   21

Faraday, Michael   20

horseshoe magnets   20

lightning   10
lights   4, 7, 17

magnets   20-21
magnifying glasses   5
mains electricity   5
microcircuits   22
Morse code   18

natural electricity   10-11

parallel circuits   7, 14
plug points   9
power stations   4
printed circuits   22-5
pylons   4

remote control   27
robot eyes   14

series circuits   6, 7
static electricity   11
switches   8-9, 14, 23, 27

wires   12, 20

WINNS PRIMARY SCHOOL
ELPHINSTONE ROAD
WALTHAMSTOW E17 5EX